Narwhals

Camilla de la Bédoyère

Quarto Library

Quarto is the authority on a wide range of topics.

Quarto educates, entertains and enriches the lives of our readers—enthusiasts and lovers of hands-on living.

www.quartoknows.com

Author: Camilla de la Bédoyère
Editor: Emily Pither
Designer: Starry Dog Books Ltd

This library edition published in 2021 by Quarto Library., an imprint of The Quarto Group.
6 Orchard Road
Suite 100
Lake Forest, CA 92630
T: +1 949 380 7510
F: +1 949 380 7575
www.QuartoKnows.com

Distributed in the United States and Canada by
Lerner Publishing Services
241 First Avenue North
Minneapolis, MN 55401 U.S.A
www.lernerbooks.com

A CIP record for this book is available from the Library of Congress.

IISBN 978 0 7112 6460 1

Manufactured in Guangdong, China TT052021

9 8 7 6 5 4 3 2 1

MIX
Paper from responsible sources
FSC www.fsc.org FSC® C016973

Photo Acknowledgments

Alamy: 7 Todd Mintz; 9 Minden Pictures; 13tr MichaelGrantWildlife, 19b Nature Picture Library, 23 All Canada Photos, 30 Nature Picture Library, 32 All Canada Photos; **FLPA:** 19t Norbert Wu, 26 Flip Nicklin; **Getty:** front cover CoreyFord, back cover, 3, 15br, wildestanimal, 1 Paul Nicklen, 11 Franco Banfi, 12-13b Paul Nicklen, 14- 15 Paul Nicklen, 20 Flip Nicklin/ Minden Pictures, 27 wildestanimal, 31 Paul Nicklen; **Marie Auger-Méthé:** 24; **Nature Picture Library:** 16 Todd Mintz, 17 Bryan and Cherry Alexander, 21 Doug Allan, 22 DOC White; **Science Photo Library:** 4-5 DAVID FLEETHAM / VW PICS; **Shutterstock:** 6 KARYI YEAP,8tl FotoReques, 8bl Paul Loewen, 8tr Vladimir Melnik, 8br Warren Metcalf, 9t Lexanda, 10 gilkop, 15tr Leonardo Gonzalez, 28 Christian Musat, 29t wildestanimal, 28-29b FloridaStock, 30t Rich Carey; **SuperStock:** 18 World History Archive, 25 Animals Animals.

Contents

What is a narwhal?

A narwhal is an animal that lives in cold Arctic seas and rivers.

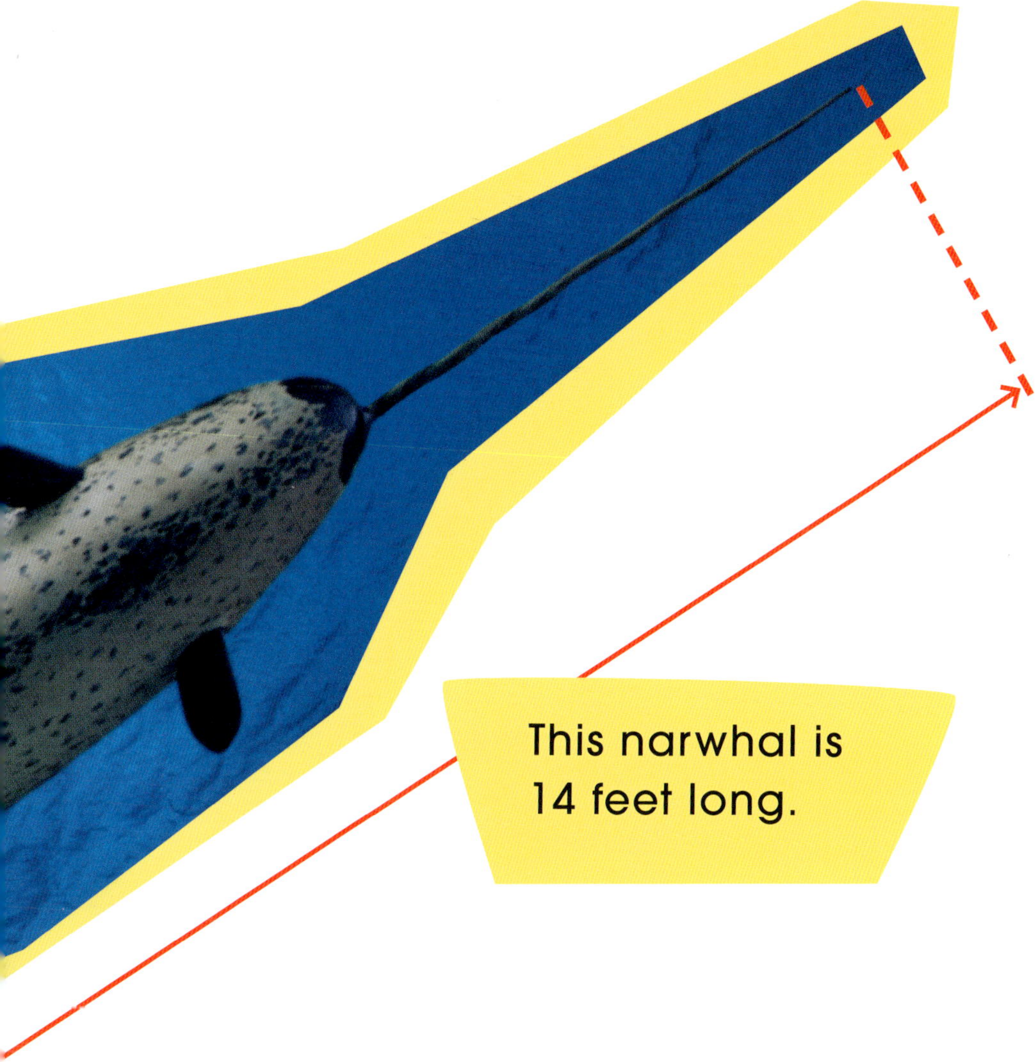

It has a long tooth that looks like a unicorn's horn.

Narwhals are called unicorns of the sea!

What do narwhals look like?

A narwhal has a long, smooth body. Its head is small and round.

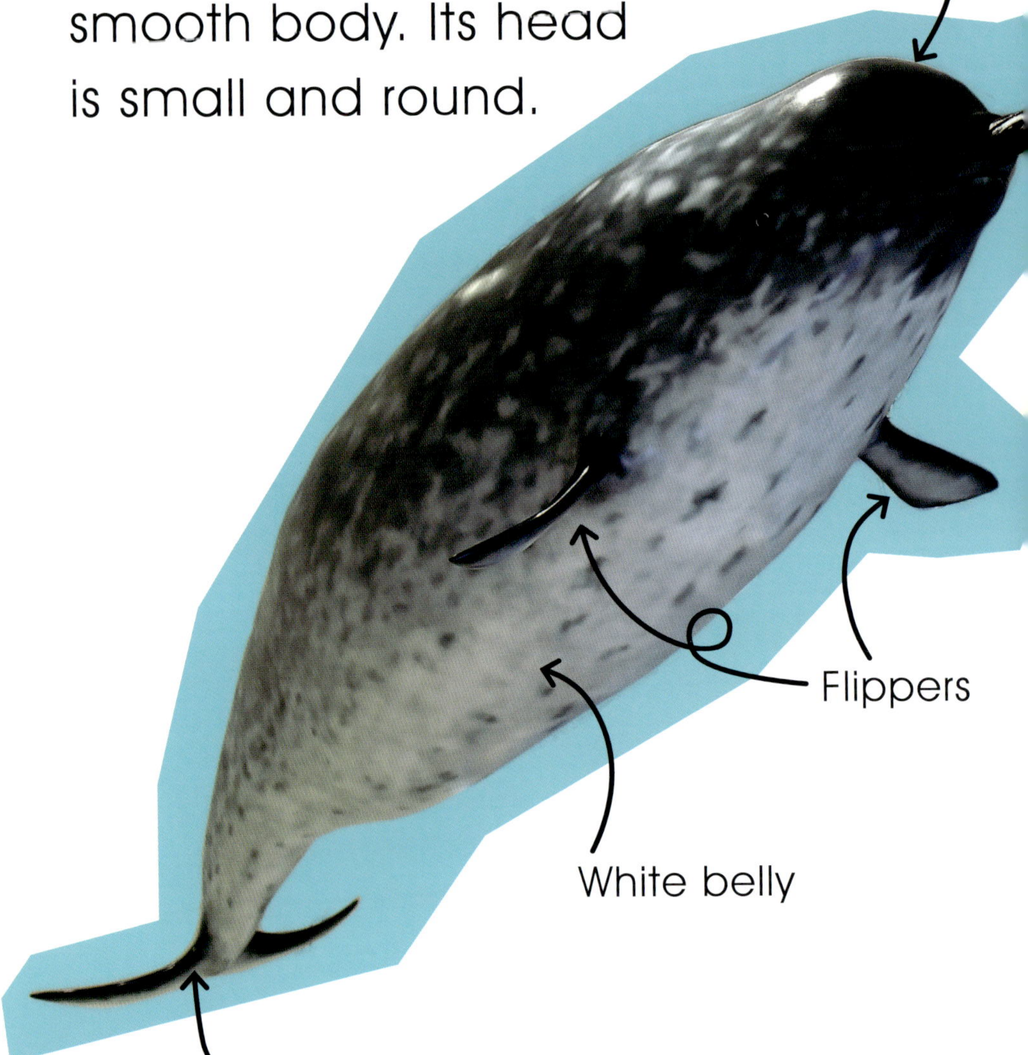

It has flippers and a tail. Its long tooth is called a **tusk**.

Cold home

Narwhals live in the Arctic Sea.

The water is very cold. Some of the water freezes and turns to ice.

These animals live in the chilly Arctic too.

This is a baby harp seal.
It is white and fluffy.

Meet the family

Narwhals are small whales.

Whales are **mammals** that live in the sea.

Dolphins are also small mammals.

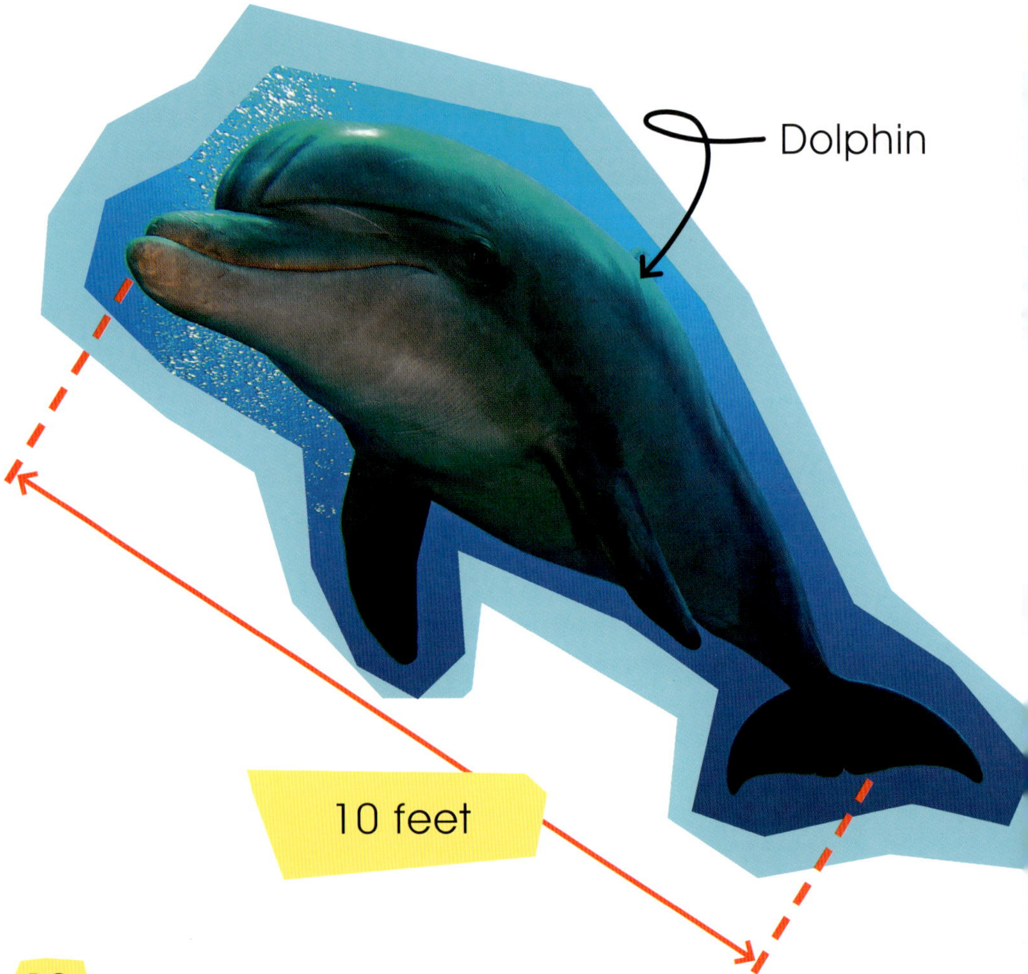

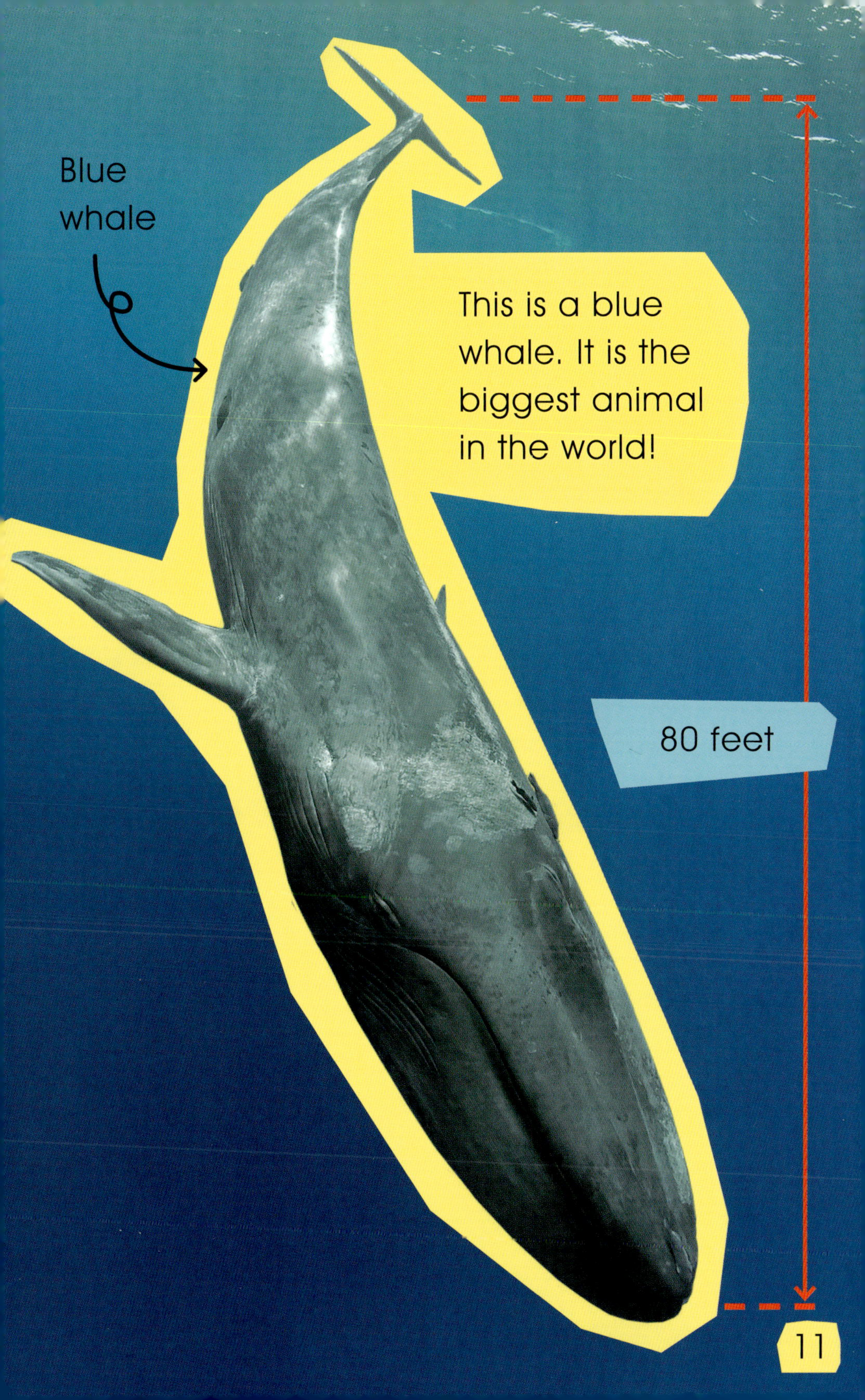

This is a blue whale. It is the biggest animal in the world!

Breathing

All whales breathe in air.

They have a hole on the top of their head. It is called a **blowhole.** Air goes into the hole.

When the whale breathes out it sends out hot, wet air!

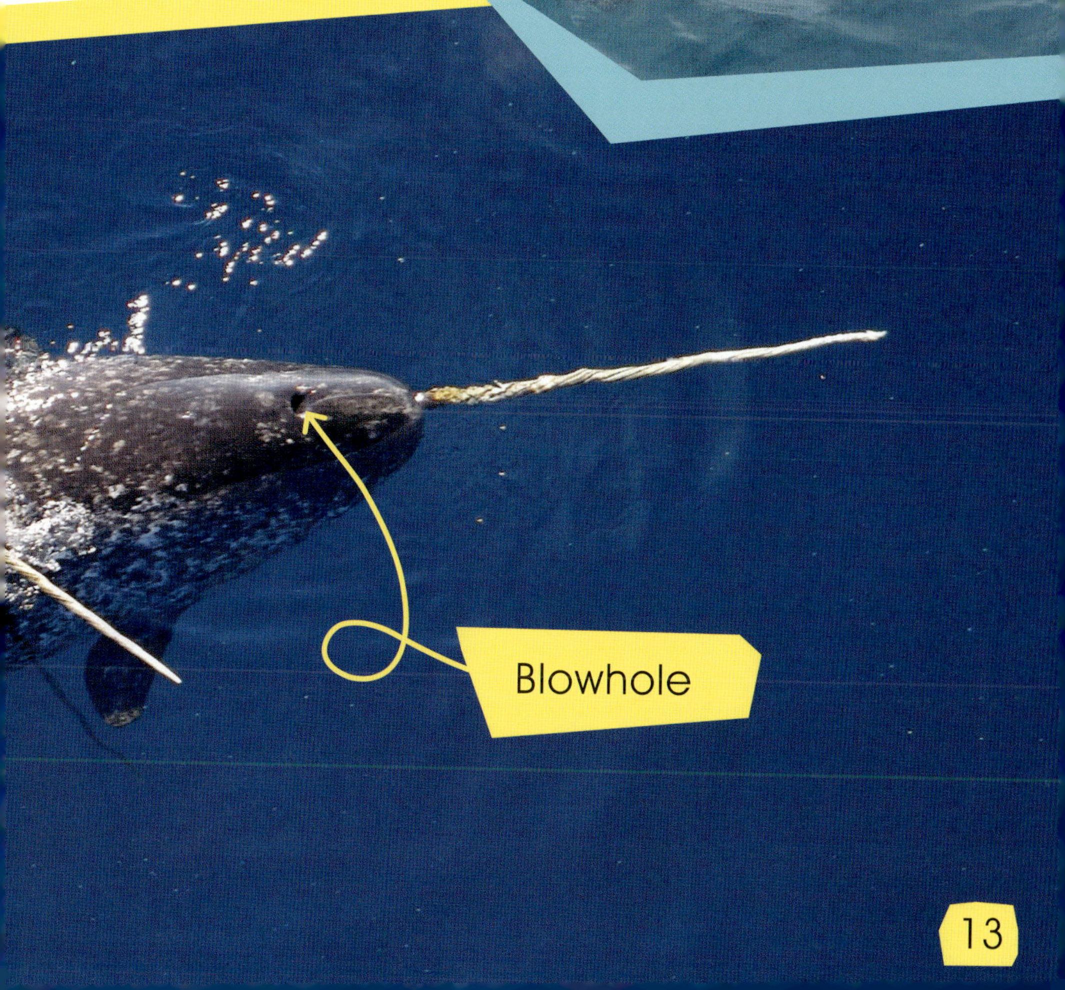

Super Swimmers

A narwhal is a good shape for swimming.

It is long and smooth so it can slip through water.

Fish and seals also have good shapes for swimming.

A narwhal uses its strong tail to push through water.

Tusks

Male narwhals have tusks. Sometimes female narwhals have tusks too.

This is a female narwhal. She does not have a tusk.

This is a male narwhal. His tusk is very long.

The tusk is long and twisted. It grows through the narwhal's lip.

What do narwhals eat?

Narwhals hunt and eat other animals. They eat fish, squid, and shrimp.

Shrimp

Narwhals dive underwater to find food. They can stay underwater for 20 minutes.

Fish

Squid

Narwhals suck food into their mouths.

Finding food

Narwhals use sound to find food.

1. The narwhal makes lots of click sounds.

2. The clicks bounce off fish that are nearby.

Using sound to find food is called **echolocation**.

3. The sounds bounce back to the narwhal.

4. It uses the sound to figure out where the food is.

What is a pod?

A family of narwhals is called a **pod**.

Most pods have up to ten narwhals in them.

Sometimes lots of pods swim together.

A very big group of narwhals is called a **herd**. There can be hundreds of narwhals in a herd!

Narwhal babies

Narwhal babies are born in the summer.

A narwhal baby is called a **calf**.

A mom feeds her calf with milk.

Narwhal babies are strong. They can swim soon after they are born.

A calf swims with its mom.

Noisy narwhals

Narwhals are noisy!

They talk to each other. Narwhals click, whistle, and squeak.

Scientists record narwhal noises.

Narwhals have small eyes. They do not see well but they have good hearing.

Narwhals use noises to find each other in the water.

Danger!

Narwhals help each other stay safe. They look out for dangerous animals.

These animals hunt narwhals for food.

Orcas are also called killer whales.

A male walrus
has two long
tusks.

Polar bears pull narwhals
onto the ice to eat them.

Saving narwhals

Narwhals are in danger because people hunt them. People also pollute the seas.

The trash in the sea harms animals.

Oil spills are harmful too.

Scientists check to make sure a narwhal is healthy.

We can help keep our planet clean and safe for narwhals, and all other animals.

Glossary

Blowhole
A hole on the top of a whale's head. It is used for breathing.

Echolocation
Using sound to find food.

Herd
A big group of whales.

Iceberg
A mountain of ice.

Mammals
Animals that have hair and feed their babies with milk.

Pod
A family of whales.

Scientist
A person who finds out about nature and the world.

Tusk
A long pointed tooth.